Soviet Mixed Power Experimental Fighter Aircraft

Piston-Liquid Propellant Rocket Engine/Piston-Ramjet/Piston-Pulsejet & Piston-Compressor Jet Engine designs of the 1940's

HUGH HARKINS

Soviet Mixed Power Experimental Fighter Aircraft

Piston-Liquid Propellant Rocket Engine/Piston-Ramjet/Piston-Pulsejet & Piston-Compressor Jet Engine Designs of the 1940's

© Hugh Harkins 2018

Centurion Publishing
United Kingdom

ISBN 10: 1-903630-76-2
ISBN 13: 978-1-903630-76-1

This volume first published in 2018

The publisher and author would like to thank all organisations and services for their assistance and contributions in the preparation of this volume: A.C. Yakovlev OKB, Central Aerodynamic Institute (TsAGI), Design Bureau of Chemical Engineering, KBhimmash, Federal State Unitary Enterprise CIAM (Central Institute of Aviation Motors), JSC Klimov, JSC Sukhoi, Ministry of Defence of the Russian Federation, NPO Energomash, NPO Saturn, Russian Aircraft Corporation, Russian Federation Air Force Museum – Monino, Scientific and Production Association, S.A. Lavochkin, United Engines Corporation

Citation guide: A.C. Yakovlev OKB (Yakovlev); Central Aerodynamic Institute (TsAGI), Design Bureau of Chemical Engineering, KBhimmash (KBhimmash), Federal State Unitary Enterprise Central Institute of Aviation Motors (CIAM); Harkins, H. (2013) 'RAF Meteor Jet Fighters in World War II, an Operational Log', Centurion Publishing (Harkins, 2013), JSC Sukhoi (Sukhoi); Ministry of Defence of the Russian Federation (MODRF), NPO Energomash (Energomash), Russian Aircraft Corporation (RAC), Russian Federation Air Force Museum – Monino (Monino) & Scientific and Production Association, S.A. Lavochkin (Laspace)

CONTENTS

INTRODUCTION

The intent of this research paper is to provide an overview of the Soviet experimental fighter aircraft programs employing mixed power plants – piston-liquid propellant rocket engine, piston-ramjet, piston-pulsejet and piston-compressor jet engine accelerator technology, in the World War II and early post war period of the 1940's. A number of piston fighter aircraft types were converted for experimental roles from the Design Bureaus of Lavochkin and Yakovlev to test liquid propellant rocket engines and ramjet accelerators to increase maximum speed of in-service fighter aircraft, Sukhoi also developing the purpose designed Su-7 as a piston-liquid propellant rocket engine powered aircraft. Lavochkin also tested pulsejet accelerators on the La-7 and La-9 piston engine fighter families whilst Mikoyan, Sukhoi and Yakovlev tested piston-compressor jet engine accelerators. The latter employed a conversion from a serial piston engine fighter whist Mikoyan and Sukhoi developed new designs for their respective piston-compressor jet engine accelerator test programs.

As no design provided the necessary combination of speed performance and reliability, the respective piston-liquid propellant rocket engine, piston-ramjet, piston-pulsejet and piston-compressor jet engine development programs, all of which were unreliable and over complex in their operation, would fall by the wayside due to the promise of better performance from the first generation exclusively jet powered fighter aircraft designs.

1

SOVIET MIXED POWER EXPERIMENTAL FIGHTER AIRCRAFT

The August 1947 air parade at Tushino, USSR (Union of Soviet Socialist Republics) included an impressive fly-past of 100 Soviet air force jet powered fighter aircraft – 50 Mikoyan MiG-9 and 50 Yakovlev Yak-15 – and a little noticed participation of mixed power fighter aircraft. While the jet powered aircraft, like their western counterparts, were far from satisfactory in overall performance advantages over the best piston engine fighter aircraft then in service, the USSR accelerated production served as a statement that her aviation industry was at least approaching par with those of her former allies, Great Britain and the United States of America. Another, more productive, reason was the provision of experience and training for pilots in preparation for the introduction of more advanced jet powered aircraft.

The MiG-9 and Yak-15 had both conducted their respective maiden flights on the same day – 24 April 1946 (pilots Alexie N. Grinchik (MiG-9) and M.I. Ivanov (Yak-15)) (MODRF). These designs featured a through engine layout with the intake in the nose, established with the early German and British jets. In the Soviet aircraft the exhaust exited through the nozzle area on the aircraft underside at roughly mid length for the MiG-9 and slightly forward of this in the Yak-15. The single crew MiG-9, developed under chief designer A. Brunov, was powered by two RD-10 turbojet engines, each rated at 900 kg (7.85 kN) thrust. The design, which included a take-off weight of 4860 kg, had a maximum speed of 910 km/h, ceiling was 13000 m and range was 900 km. Armament consisted of two HC-23 23 mm cannon and a single H-37 37 mm cannon (RAC & Monino). The Yak-15 was notably smaller than the MiG-9, with a normal flight weight of 2638 kg. The airframe was developed from the Yak-3 piston engine fighter aircraft airframe with deep modifications to allow the incorporation of a single RD-10 turbojet engine rated at 900 kg thrust. The low power output, despite the fact that the aircraft was lighter than the MiG-9, resulted in overall performance being somewhat inferior to that of the MiG aircraft, maximum speed being put at 805 km/h. The wing conformed to the straight wing (area was the Yak-3 14.85 m^2) concept of the early jet aircraft of the western powers.

Figure 1. The Su-5 was developed in the last year of World War II as a mixed power fighter aircraft - piston engine integrated with a Compressor Jet Engine. Sukhoi

Figure 2. In regards to speed the MiG-9 (I-300) was superior to the jet powered Yak-15, piston only powered fighters and mixed power fighters available in 1946. MODRF

The Yak-15 went on to become the first Soviet jet powered fighter aircraft to enter service, production, which commenced on 5 October 1946, extending to 280 units built during 1946 and 1947 (MODRF & Yakovlev). Several hundred MiG-9 fighters were also built, providing a viable early post-war embryonic Soviet jet fighter force as more advanced designs were developed. A development of the Yak-15, the Yak-17UTI two-seat conversion trainer aircraft, was produced serially, 430 units, powered by the RD-10, being built during 1948 and 1949. The 2806 kg weight Yak-17UTI had a reduced maximum design speed of 725 km/h. This, however, was considered acceptable for the conversion trainer role (Yakovlev).

The MiG-9, Yak-15 and Yak-17UTI provided the Soviet air forces with an embryonic operational capability with jet powered aircraft. Later Soviet fighters in the mold of the MiG-15 would continue the nose intake layout, but with the engine positioned in the aft section of the aircraft and the exhaust exiting the extreme rear, building on experience gained, after a fashion, with the BI-1 rocket powered experimental interceptor and the piston-rocket and piston-CJE (Compressor Jet Engine) mixed power fighter designs of the mid-late 1940's.

Figure 3. The German Me.262 twin-jet fighter, which entered service in summer 1944, was regarded as the benchmark of late war performance that had to be aspired to – maximum speed being in the region of 850 km/h under normal conditions. NMUSAF

Development of the Soviet mixed power designs had been initiated during the war years as a means of improving the performance of in-service fighter aircraft. In the last year of the second world war jet and rocket powered aircraft had begun to make their mark on the world stage – manned jet and rocket powered aircraft were in operational service with Britain and Germany. The Gloster Meteor MK. I//III twin-jet fighters (the prototype F.9/40 had conducted its maiden flight on 5 March 1943) entered service in July 1944 and March 1945 respectively. The Messerschmitt Me.262 twin jet fighter, the first jet powered flight of which took place on 25 March 1942, entered service in summer 1944. The Arado 234 twin jet bomber flew in 1943 and entered service in late summer 1944. The Messerschmitt Me.163 single-seat liquid propellant rocket engine powered interceptor flew under the power of a HWK 109-509A rocket engine in July 1943 and entered service in summer 1944. The maximum flight speed values for the Me.262 and Me.163, at 850-878 km/h for the former and 940 km/h for the latter, were considerably greater than those of the British Meteor I – 410 mph (~660 km/h) at sea level and 430 mph (~692 km/h) at 10,000 ft. (~3048 m) and the Meteor III – 476 mph (~766 km/h) at 10,000 ft. rising to 484 mph (~779 km/h) at 30,000 ft. (~9144 m) when powered by two W.2B/37 units each rated at 2,000 lb. (~907 kg) static thrust. In the immediate post war period the Derwent 1 powered Meteor, when producing 2400 lb. (~1089 kg) thrust, was credited with a speed of 505 mph (~813 km/h) at sea level and 520 mph (~837 km/h) at 30,000 ft. The Meteor IV, which flew in August 1945, was capable of 585 mph (~941 km/h) at 30,000 ft. (Harkins, 2013). A flight test report dated 1 September 1944 credited the Me.262 with a speed of 820 km/h at 9000 m. A Messerschmitt report dated 15 October 1944 credited the Me.262 with a speed of 850 km/h at altitudes between 6000-8000 m, normal operating speed being somewhere in between (Harkins, 2013). Britain and Germany were also experimenting with or developing for service a number of other jet powered aircraft and jet powered fighters were being developed in the United States of America. The single-seat, single jet engine, de Havilland Vampire MK. I, which was capable of 540 mph (~869 km/h) at 34,000 ft. (~10363 m) entered British service in 1946 (Harkins, 2013). The Soviet Union, fighting a different type of war from that of the western powers – a war of survival until the tide turned against Germany with the defeat at Stalingrad when the Soviet effort became a war of liberation of occupied lands – was less occupied during the war years with jet aircraft development.

Figure 4. The Derwent powered Gloster Meteor III entered service in March 1945. CC

With the introduction to service of jet and rocket powered aircraft in 1944, the Soviet mixed power fighter programs took on a new importance, alongside jet powered fighter aircraft development, to ensure Soviet air power was not to be outclassed going into the second half of the 1940's. During the war this development effort was conducted without the benefit of German technology data. However, Soviet Jet propulsion and rocket powered research, as it was for the western allies, was considerably expanded at the end of the war with the availability of design data from captured German material, records and personal.

For decades it had been considered in the western orbit that the Soviet Union jet/rocket power for aircraft programs had commenced toward the end of the European war in May 1945 as a reaction to British and German developments. However, this is a fallacy, Soviet research and development into alternative power plants for aircraft pre-dating the start of the Great Patriotic War, which had commenced with the German invasion of Soviet territory on 22 June 1941. The commencement of war highlighted the need for accelerating development of jet and rocket powered aircraft whilst at the same time installing a climate of reduced capacity for development of such programs, which could not be allowed get in the way of the immediate war aims that required maximum output of existing weapons and introduction of near term weapon programs, all in a climate of upheaval as many bureau and plants were evacuated in the face of the German advance.

In the last year of the war, 1944-1945, the two major priorities for Soviet fighter aircraft development centred on increasing range to provision for an escort fighter to accompany projected long-range bomber developments and increasing maximum speed. As jet power for aircraft was in its infancy, the first generation turbojet powered aircraft having poor range performance compared to piston engine aircraft, the first of these requirements would be met with an evolution of the Lavochkin piston engine fighter family, resulting in the La-11. The second requirement, it was clear, would have to be met with either jet or rocket propulsion or mixed power (piston engine augmented by rocket or jet accelerators) building on research and development work that had been underway in the Soviet Union for some time.

Figure 5. The Soviet late-war requirement for a long-range escort fighter was met by the Lavochkin La-11 piston engine fighter, which entered service in the immediate post-war years. Laspace

In the years before the first Soviet jet/rocket powered aircraft took to the air, Soviet designers had conducted research and development on rocket propulsion for aircraft and various types of jet propulsion as both main power plants or as booster/accelerators to increase the performance of piston engine aircraft. In Germany, Physicist Dr. Hans von Ohain was preparing a jet engine for flight. Ohain's vision of a jet (gas turbine) engine emerged later than Britain's Frank Whittles idea for such an engine in Britain. However, Ohain was fortunate that his bureaucratic masters in Germany were more enthusiastic about the prospect of jet powered high-speed flight than their counterparts in Britain. Consequently, Ohain received the funding and facilities he required to turn the idea into reality, leading to the Heinkel He.178 becoming the world's first aircraft to fly exclusively under jet power when it conducted its maiden flight on 27 August 1939 powered by a single He S-3b engine rated at 1,100 lb. (~499 kg) static thrust. Whittles vision was realised when the Gloster E.28/39, powered by a single W-1 (Whittle-1) gas turbine engine, took to the air on 15 May 1941 (Harkins, 2013). Although largely overshadowed by the volumes of data available on the British and German jet developments, led by Whittle and Ohain, in the years before the outbreak of World War II in September 1939, in the Soviet Union the academician B.S. Stechkin had developed the basic principles of jet engine function as early as 1929 (CIAM). The entity for the organization in the GDL (Gas Dynamic Laboratory) was established in Leningrad in 1929 and on 15 May that year, V.P. Glushko was invited to commence practical research and development into rocket engines under the auspices of the GDL. This work would build on Glushko's research into 'Metal as explosive' used to graduate Leningrad University. This resulted in a team, led by Glushko, designing the world's first electro-thermal rocket engine, a patent being submitted and accepted for this work by 1930. From this point Glushko specialised in the development of what would become known as LPRE (Liquid Propulsion Rocket Engines) (Energomash).

In 1930-1931, Glushko's team at the GDL created the experimental liquid propellant rocket engines ORM-1 (this was the first Soviet designed truly practical LPRE, which was ready by summer 1931) and ORM-2, 47 bench test runs of these engines being conducted in the latter year. In this period the GDL team also studied various innovations such as the 'cardan mount' for 'engine with pump units'. Studies into substances to feed the rocket engines included, among others, the use of nitric acid as fuel (Energomash). Experience gained in development of the ORM-1 and ORM-2 fed into the design work on the ORM-4 to ORM-22 designs of 1932. Such engines featured an ignition system that involved 'start-up and systems of mixture' through the use of different propellant components, an early form of chemical ignition (Energomash). Continued development led to the ORM-23 up to ORM-53, which featured a system of 'pyrotechnic and chemical ignition', which were developed and tested in 1933 (Energomash). Going into 1934, the ORM-53 was still being developed and in the years to, and including, 1938, rocket designs up to the ORM-102 were developed. This included the ORM-65 LPRE that would power the RP-318-1 rocket glider and the 212 cruise missile design (Energomash).

In 1934, work on developing rocket power plants for aircraft had commenced at KBhimmash (KB V.F. Bolkhovitinov) as a joint effort of A.M. Isaeva and A. Ya. Bereznyak with the specific aim of developing a rocket powered interceptor (later referred to as BI (Berezyak-Isayev) (KBhimmash). When it conducted its maiden flight on 15 May 1942, the BI-1 was regarded as the first practicable Soviet designed rocket powered aircraft to fly. The Korolev RP-318 can, perhaps, be more accurately categorised as a low speed air launched object to test primitive LRPE technology rather than as an aircraft in the sense of take-off, flight, and landing. For practical military purposes the BI-1 is considered by the former Soviet Defence Ministry and the Ministry of Defence of the Russian Federation as the first true practical alternative power – rocket or jet propelled – aircraft to fly in the Soviet Union. The pilot for the BI-1 maiden flight was Captain G.Y. Bahchivandzhi. Bahchivandzhi, who had been taken from operational flying, having flown in the defence of Moscow in winter 1941, to test fly the new rocket powered aircraft, was killed in an accident on the BI-1's seventh test flight on 27 March 1943 (MODRF & KBhimmash). Initially to be powered by a D-1-A-1100 rocket engine, later aircraft were powered by rocket developments underway at Energomash and KBhimmash. The resultant RD-1 (designed by Glushko) developments were born out of early Soviet pioneering work into rocker propulsion conducted at OKB-456 (NPO Energomash).

Figure 6. The RP-318 rocket powered glide vehicle of the late 1930's.

Figure 7. The prototype BI-1 experimental rocket powered aircraft. KBhimmash

With onset of Second World War the development of LPRE to power aircraft had been accelerated in the Soviet Union, leading to not only the RD-1 and the RD-1KhZ (also referred to as XZ) – both units had a thrust rating of 300 kgf. The RD-2 had a thrust rating of 600 kgf and the experimental three-chamber RD-3 had a thrust rating of 900 kgf (Energomash). Isaeva was appointed head of engine development at KBhimmash on 21 June 1943. That same year saw the development (and further development) of RD1, RD1-M, U-1250, SU-1500 and U-400-10 LPRE (KBhimmash). On 7 December 1944, an experimental design bureau, OKB-SD, was formed within OKB-436 with the specific aim of developing LPRE's for aircraft. OKB-SD was moved to Khimki in the Moscow region on 3 July 1946 (NPO Energomash). During the war years and it's near term aftermath the various Soviet LPRE designs were flown on Petliakov (Petlyakov) (V.M.) Pe-2R, Lavochkin La-7R, Lavochkin 120R, Yakovlev Yak-3, Sukhoi Su-6 and Su-7. In excess of 400 engine start-ups were conducted on these aircraft. Small-scale serial production of the RD-1 commenced in 1944, followed by small-scale serial production of the RD-1KhZ in 1945 (Energomash). Despite this small-scale production of the RD-1 variants, LPRE as accelerators for aircraft was gradually proving unnecessary due to the promise of jet propelled aircraft in development as the war was coming to a close in summer 1945. Although testing with the RD-1 variants continued for a short time after, the direction of work on LPRE for OKB-SD was now focused on more powerful engines for ground launched rockets. To this end a Soviet team from OKB-SD was dispatched to Germany in June 1945 to garner data and materials on the A-4 (FAU-2) rocket, more commonly referred to as the V-2, which would be further developed in the Soviet Union as the R-1 short-range ballistic missile (Energomash).

In 1946 the development effort on the BI-1 was brought to a close, attention turning to turbojet powered interceptor/fighter aircraft (KBhimmash). During ~four years of testing the BI-1 had attained a maximum speed of ~800 km/h, proving to be considerably inferior to the German Me.163 Komet (this aircraft was capable of 940 km/h) that had entered operational service in 1944. TheBI-1 performance was short of that required for a viable interceptor design, the program taking on the shape of a dedicated research effort to further the development of rocket propulsion. With the end of the BI-1 development rocket engine developer KBhimmash moved away from development of engines intended for aircraft and concentrated on LPRE for missile's, in particular anti-aircraft missiles and the 'Storm' cruise missile developed by Lavochkin (KBhimmash). Rocket powered aircraft design did not end however, the Mikoyan I-270 flying in December 1946. This program would not result in a viable fighter/interceptor suitable for service, the LPRE being left in the shadow of turbojet engines of increasing power that were now promising significant performance increases over not only the best piston engine fighter aircraft, but also the early jet powered fighter aircraft.

Figure 8. A BI-1 experimental aircraft preserved at Monino, Moscow. Monino

MIXED POWER FIGHTER DESIGNS - While research and development work was ongoing into rocket powered interceptor developments, V.V. Uvarov, who had conducted work on gas turbine engine research since 1930, was, in 1943, working on combined piston-jet exhaust engine research at CIAM (Central Institute of Aviation Motors). This research would feed into a number of the mixed power developments that were flown in the period 1945-47 (CIAM). In the same year that BI-1 conducted its maiden flight, a research effort within OBK Lavochkin was accelerated to provide jet accelerators to increase the performance of existing piston engine fighter aircraft.

Figure 9. Lavochkin LaGG-3 piston engine fighter aircraft. TsAGI

The first such development was the Lavochkin LaGG-3VRD, which was a LaGG-3 piston engine fighter aircraft equipped with two VRD-1 ramjet accelerators, which were attached by brackets to the underside of the wings near the second rib section. The LaGG-3 was itself developed from the LaGG-1 single-seat piston engine fighter with the specific aim of extending the range of the latter. Development of the LaGG-3 had commenced in January 1941 and production was organised for Plant No.21 in Gorky. The production schedule called for the manufacture of 2,960 LaGG-3 fighters during 1941, only 322 of which had been delivered by 22 June 1941, the date Germany and her allies launched Operation Barbarossa, the invasion of Soviet territory. The LaGG-3 had many faults, the details of which are beyond the scope of this paper. Its performance was inferior to the German Messerschmitt Me.109F single-seat fighter and the Soviet Yakovlev Yak-1 single-seat fighter. To improve flight performance S.A. Lavochkin and V.P. Gorbunov embarked upon a series of design improvements that included installation of a new piston engine – the M-105PF. Further improvements led to the creation of the La-5 single-seat fighter, the LaGG-3 itself having performance improvements that allowed it to fall not too far short of that achieved by the Yak-1 and Yak-7.

Figure 10. This poor quality photograph apparently shows the LaGG-3VRD on approach for landing, although the VRD-1 accelerators appear to be absent. Laspace

The LaGG-3 design has received much criticism down the decades since it operated during World War II. While much of this criticism is indeed warranted, the aircraft being clearly inferior in performance to the best foreign and domestic Soviet fighters then in existence, it has to be considered that the poor performance of many Soviet air units in the early part of the war – 1941-1942 – was due, at least in part, to the low quality of pilot training. Average Soviet pilots had 30 hours training before entering combat on the LaGG-3 whilst their Luftwaffe opponents, many of whom had considerable experience of operations over the European mainland and Great Britain in the campaigns of 1939-1941, had on average 450 hours. The LaGG-3 played an important part in the early battles with the Luftwaffe, particularly in the defence of Moscow in October 1941, inflicting considerable losses on the Luftwaffe, but suffering considerable losses of their own in the process. Throughout the war the best successes were achieved against enemy bomber aircraft, the LaGG-3's heavy armament proving very effective. Against German fighter aircraft the overweight LaGG-3's poor maneuverability and overall inferior performance resulted in fewer successes. The poor performance against the Me.109F/G led to production being switched to other fighter designs – La-5 (Plant No.21 at Gorky) and Yak-7 (at Novosibirsk) except at the Tbilisi plant (this was established from the Taganrog aircraft plant No.31, which had been evacuated in the face of the German advance into the Soviet Union. Total production of the LaGG-3 at all plants, including Plant No.23 (Leningrad), amounted to 6,528 aircraft when production ceased in 1944.

The LaGG-3 basic performance included a ceiling of 9000 m and maximum speed values of ~555 km/h (Monino). In an attempt to address the LaGG-3 performance shortfalls in the area of speed a single aircraft was taken in hand to be modified as the LaGG-3VRD ramjet testbed. In this configuration the aircraft conducted its maiden flight on 5 August 1942, almost three months after the maiden flight of the BI-1 rocket powered experimental interceptor. A total of 13 flights were conducted, the last of which was flown on 15 August 1942. In comparison to the standard serial LaGG-3 a modest increase of 12-15 km/h in maximum speed was achieved. The small scale of flight tests had shown that considerable research and development would have to be conducted in order to achieve acceptable performance improvements that would justify the detrimental effects of being encumbered by the weight of two ramjets. In this respect, the high workload of Lavochkin dictated that such work on ramjet accelerators would be all but suspended until the end of the war against Germany (Laspace).

In the nearer term the use of LPRE to increase maximum speed of in service fighter designs held more promise. As briefly noted above, several aircraft types would fly with LPRE installed under the Soviet development program(s). One such platform was the twin engine Pe.2, which was developed as a high-speed twin-engine (2 x M-105PF piston engines each rated at 1260 hp.) dive bomber for front line aviation. The Pe.2, which entered serial production in 1940, had a design maximum speed of 540 km/h, ceiling was 9000 m and the aircraft was armed with 3 x 12.7 mm machine guns, one 7.62 mm machine gun and a bomb load of 600-1000 kg. The single Pe.2R (distinct form the reconnaissance variant) was equipped with the RD-1 LPRE, not to increase the performance of the aircraft itself, but as a flying laboratory to flight test LPRE intended for installation on other platforms, including the mixed power fighter designs under development in the last years of the war.

At Lavochkin design bureau plans to improve the speed performance of the La-7 single-engine-fighter resulted in the La-7R(P), which was an adaption of the serial La-7 piston engine fighter, adding a single RD-1 LPRE as an accelerator for an emergency. The La-7, initially known as the improved La-5 fighter 'Standard 1944', was an evolution of the La-5FN based on a series of recommendations from TsAGI (Central Aerodynamic Institute). The new design was not merely an update of the La-5FN, but closer to a new design altogether. There were significant changes introduced, such as metal wing spars. The AH-82FN piston powered La-7 had overall superior performance to its forebear in the areas of climb rate and in the non-flight area of armament (initially the La-7 adopted the same armament as that of the La-5FN, but in 1945 this was changed to three 20 mm Berezina UB-20 cannons, each with a capacity for 135 rounds). Both designs had a standard operating ceiling up to 11000 m and speed values were similar. The La-5 had excellent speed performance values for its time – 650 km/h and the La-7 standard maximum speed value was, at 644 km/h, close to that of the La-5. This was achieved at a lower operational weight than that of the La-5FN.

The La-7 prototype had conducted its maiden flight in January 1944 (LII test pilot G.M. Shiyanov) and the first series production examples were rolled out of the factory in May and entered operational service with Soviet aviation units in June that year. During the course of 1944-1945 a total of 6,209 La-7 aircraft were built at Plant No.21, 99 and 381 (Laspace). It is a point of historical fact that the first air combat in which a Soviet fighter destroyed a German jet aircraft in the air occurred on 24 February 1945 when a La-7 pilot of the Soviet 2nd Air Army (often mistaken as a Yak-9 unit) shot down a Me.262 twin-jet fighter near the Oder River (MODRF).

Figure 11. The La-7 was an evolution of the Lavochkin La-5FN single-engine fighter. TsAGI

Figure 12. Lavochkin La-7 single-engine fighter. Laspace

In a drive to improve the performance of the La-7 further one aircraft was retrofitted with an M-71 (AS-710) piston engine, but the test program was never completed as development work on the engine was cancelled (Laspace). A more radical attempt to improve the performance of the La-7 to enable it to close the speed gap with the German jet fighters was the La-7R(P), noted above. This added an RD-1 LPRE booster to the existing piston engine power plant.

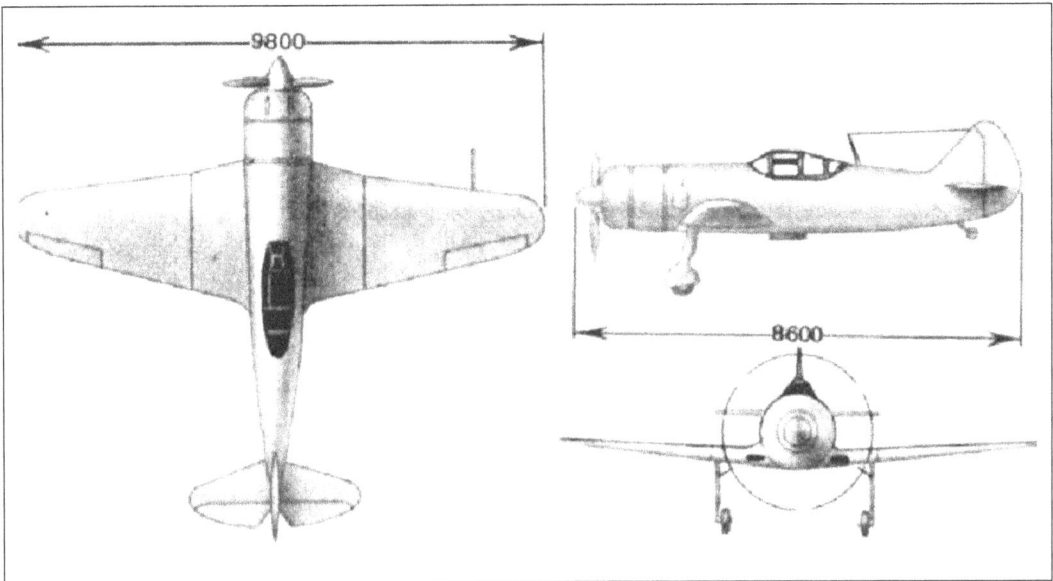

Figure 13. Three-view general arrangement of the La-7 without the LPRE. (Monino)

Figure 14. The La-7R circa 1944/45. Laspace

The first La-7R fighter was built at Plant No.81 (this was located on the grounds of the Moscow serial Plant No.381) in October 1944. This aircraft was, under the direction of S.M. Alekseeva, modified to house the RD-1 LPRE in the rear fuselage, exiting through the lower tail section, to augment the power of the AH-82FN piston engine in the nose. A pumping mechanism allowed the transfer of fuel components of the AH-82FN engine to the RD-1, the latter burning kerosene and nitric acid. Two La-7R aircraft were built and flight tested from 27 October 1944 until 24 February 1945. Test pilots for these flights were G.M. Shiyanov and A.V. Davydov.

In regards to performance improvements the installation of the RD-1 accelerator increased maximum speed by 90-100 km/h. Although this was a significant increase over the piston only speed of the La-7, the accelerator proved to be unreliable and over complex in its operation. It was decided that further research work was required in the field of LPRE, the La-7R not progressing to serial production.

While work was been underway on the La-7R program another research program commenced utilising the Lavochkin 120 development aircraft fitted with the RD-1KhZ LPRE, the 120 carrying the 120R designation post installation of the rocket accelerator. Much of the work on the La-7R and 120R programs was conducted simultaneously as both had the same end employing different platforms. The 120 had emerged as a further improvement of the La-7 of which it was more or less externally identical. In the fuselage structure the major difference over the La-7 was the adoption of metal structure in the forward fuselage, the rear fuselage retaining the wooden structure of the La-7. The major design change concerned the wing, the 120 adopting a new 'laminar profile' designed to increase maximum speed. However, the major driver in increasing performance over the serial La-7 was the adoption of the 1900 hp. ASh-83 (AH-83) piston engine, which passed bench tests in 1944. In August 1944 a serial La-7 fighter aircraft powered by an ASh-83 exceeded 700 km/h (435 mph), putting it in a similar performance class to the best western allied and German piston engine fighter aircraft. This paved the way for its adoption on the 120, which in effect became a prototype design that would lead to the La-9.

Figure 15. Poor quality photograph of the Lavochkin 120/R. Laspace

The first of two Lavochkin 120 aircraft conducted its maiden flight on 2 January 1945 (pilot A.V. Davydov). Flight tests showed a significant speed performance increases over the serial La-7 – a speed of 725 km/h (450.5 mph) being achieved in July 1945. This was an increase of 24 km/h over the maximum speed of the La-7 powered by the ASh-83 piston engine.

Work also continued on furthering development of platforms for incorporation of LPRE. As it was at least 24 km/h faster than the ASh-83 piston engine powered La-7, it was decided to utilise the Lavochkin 120 powered by an ASh-83 piston engine. A new development of the RD-1 accelerator had been developed by Glushko. This, the RD-1KhZ noted above, featuring true chemical ignition, was fitted to the Lavochkin 120 aircraft, requiring considerable modifications to the aircraft, particularly the tail section where the rocket exhaust exited. In this new configuration the aircraft, as noted above, was designated 120R.

The maiden flight of the 120R took place in July 1945 (pilot A.V. Davydov), flight and ground testing continuing into August 1946. Lavochkin documentation states that a speed increase of 103 km/h was attained over the unmodified 120 speed. Despite the increase in speed, as had been the case with the La-7R, it was deemed that the LPRE use as accelerators for existing piston engine fighter aircraft was unreliable and work on the program was discontinued. However, the 120R was cleared for public display and participated in the 1946 Tushino air parade.

While Lavochkin was progressing with the La-7R and 120R developments, the use of rocket accelerators for piston engine fighter aircraft was also being applied to non-Lavochkin designs, specifically the Su-7 and the Yak-3. While the latter was an adaption of an in-service piston engine design, the former was developed specifically as a mixed power fighter, there being no piston engine variant in-service or, indeed, planned.

Figure 16. A Sukhoi Su-6 development aircraft without the RD-1KhZ modification. It is unclear what model of Su-6 was converted as LPRE testbed Sukhoi

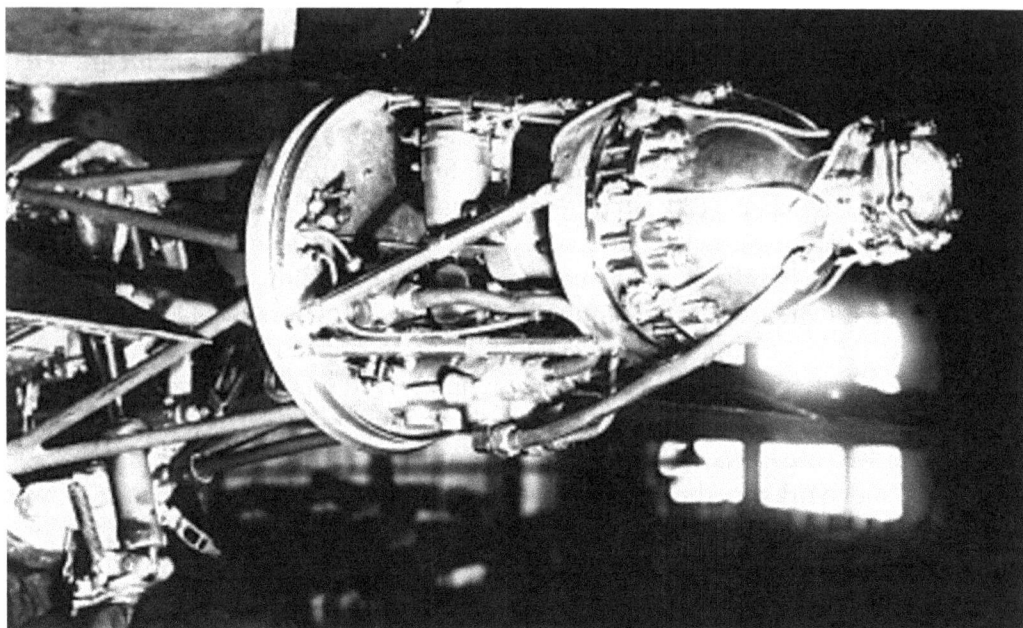

Figure 17. RD-1KhZ LPRE variant fitted in the tail section of the Su-7. Sukhoi

Figure 18. The Sukhoi Su-7 piston/LPRE mixed power experimental fighter. Sukhoi

As briefly noted above, it is known through NPO Energomash documentation that at least one Sukhoi Su-6, initially developed as an armoured ground attack aircraft in competition with the Ilyushin IL-2, was outfitted as a LPRE testbed. This was probably in support of the Sukhoi Su-7 mixed power, piston/LPRE experimental fighter development. The Su-7, initially known under the Sukhoi Design Bureau code '82 G', was powered by a single ASh-82FN piston engine positioned in the aircraft nose section. An RD-1 variant (it is unclear if this was an

RD-1 or an RD-1KhZ, but probably the former as Yakovlev was allocated priority for the more reliable LPRE for its Yak-3 RD-1 modification) rocket accelerator was positioned in the tail section. The Su-7 ('82 G') conducted its maiden flight (pilot G.I. Kamorov) on 11 January 1945 and completed factory tests in November that year (Sukhoi). Like previous piston engine fighter designs with rocket accelerators, the Su-7 design was considered inadequate in performance, encumbered by the weight of the rocket accelerator, which, as had been the case with the Lavochkin piston engine/LPRE designs, proved unreliable in operation.

The Yak-3 (second incarnation of this designation) was a single-seat monoplane fighter aircraft powered by the same M-105 пф piston engine as installed in the Yak-9. Although adopting the same power plant as that of the Yak-9, the Yak-3, at 2660 kg, was 400 kg lighter than its forebear, with a newly designed wing with an area of 14.85 m² compared with the 17.25 m² of the Yak-9. The Yak-3 had excellent performance for its time, including a maximum speed of 655 km/h. The first examples reached operational units in summer 1943 and took part in the Battle of Kursk, proving superior to the German Me.109 variants and the FW.190. In early 1944 plans were formulated for a conversion of the Yak-3, retaining the M-105 пф piston engine, which would now be augmented by a single RD-1 variant LPRE housed in the aircraft tail section. The installation of the RD-1 on the Yak-3, as was the case in other design conversions, was intended to provide a workable increase in speed in the event of emergency such as entering combat with enemy jet powered aircraft in the class of the Me.262 twin-jet fighter that had entered service in summer 1944, news of which had reached the Soviet leadership. Work on the Yak-3 RD-1 conversion geared up in December 1944 and the design conducted its maiden flight as a piston/LPRE accelerator test-aircraft in early 1945. A total of 21 flights were apparently flown. The conversion apparently attained a speed of 782 km/h (~485 mph). Despite claims that speed was increased by in excess of 182 km/h over the unmodified Yak-3, the speed of 782 km/h would correspond to a true speed increase value of 127 km/h (79 mph) when the maximum speed of 655 km/h of the unmodified aircraft is used as the start point.

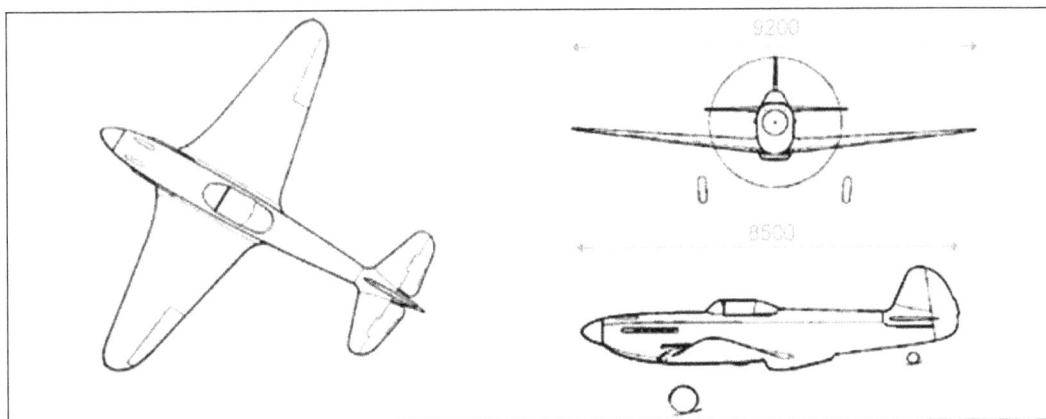

Figure 19. The serial piston engine Yak-3 layout remained relevant to the RD-1 conversion with the exception of the modifications to the rear section. Yakovlev

Figure 20. Unmodified serial Yak-7 fighter. MODRF

The Yak-7 was a single-seat monoplane fighter aircraft powered by a single piston engine. The design, which had a wing-area of 17.5 m², a flight weight of 3010 kg and a maximum speed of 613 km/h, entered serial production in 1941. Total production amounted to 6,120 units (Yakovlev). As can best be ascertained from available data there were two separate configurations of Yak-7 mixed power fighters equipped with ramjet accelerators. The first modification employed the ramjet in conjunction with a LPRE accelerator to boost maximum speed. The second modification was equipped with two ramjet accelerators to boost speed of the aircraft attained on the piston engine.

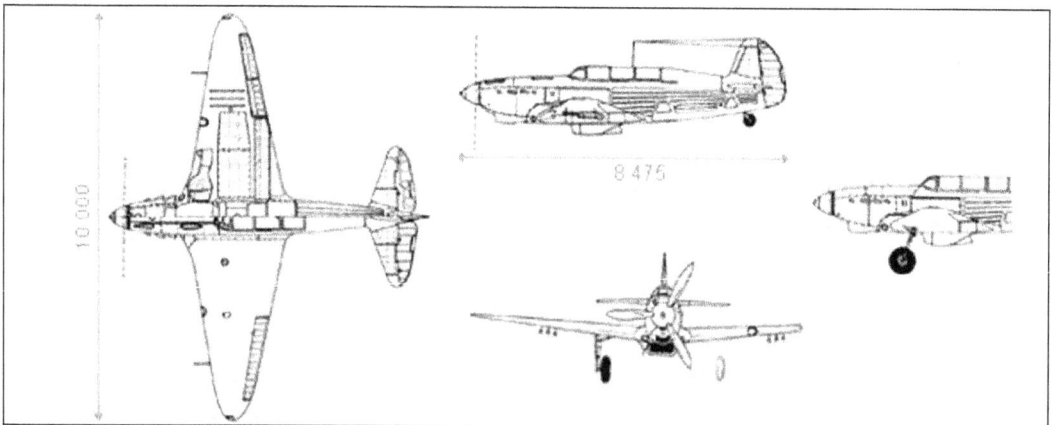

Figure 21. General arrangement of the unmodified Yak-7 fighter of 1942. Yakovlev

Figure 22. I.153bis bi-plane fighter aircraft equipped with ramjet accelerators.

There is sparse reliable data available concerning the summer 1942 Yak-7 conversion with the DM-4 (ramjet)/DM-1 (D-1A) LPRE. Specifically it has not been possible, at the time going to press, to have official 'agency or bureau' confirmation of whether or not the conversion was to fly as a piston/ramjet/LPRE demonstrator or as a ramjet/LPRE demonstrator only. The available data is conflicting. One configuration put forward concerns conversion with the ramjet sustainers mounted on the wing undersides and a LPRE accelerator in tail section, but the propeller installation still present in the nose, strongly suggesting the presence of the piston engine. Again, removal of the piston engine would have seriously affected the aircraft centre of gravity and overall stability in flight unless a substitute weight was positioned in the engines stead. If, indeed, the design was to be flown as a ramjet/LPRE testbed without piston power then this could have been done with the engine feathered, but the propeller system, of course, would provide unnecessary drag with the corresponding detrimental effect on speed performance. It is suggested that a cowing would replace the propeller hub. Considering the raison d'être of the conversion was to increase the overall speed performance of the Yak-7 fighter in the face of German high speed aircraft developments, it would be prudent to conclude that the such a design with no piston power would have been all but useless as anything other than a point defence interceptor, this fact probably having prominence in the decision to abandon the project before the aircraft had flown.

Several years prior to installation on the Yak-7, DM boosters had been flight tested on the I-15bis and I-153bis single-seat, single piston engine, fighter aircraft. The flights on these aircraft were aimed primarily at testing the boosters in a flight environment. The I-15bis apparently attained a speed increase of just over 20 km/h over the unmodified aircraft on piston only power. The I-153bis apparently attained a speed increase of just over 40 km/h over the unmodified aircraft on piston only power. It should be noted that, despite suggestions to the contrary, the I-15 and I-

153bis conversions were basically intended as test-beds to flight test the DM series ramjets and were not in themselves intended to evaluate the engines for any state sanctioned serial production variant of these aircraft.

In early 1944, a Yak-7 serial production aircraft was modified to carry two DM-4 ramjet accelerators (designed by I.A Merulov), one under each wing as accelerators to boost flight speed in an emergency. Flight testing, under piston only power, commenced in late March 1944 and the first flight on which the ramjets were operated took place on 15 May that year (the test pilot for this flight was apparently S.N. Anokhin (LII)). The results of the flight test phase concluded that a speed increase of only 52-53 km/h was attained over the unmodified piston only aircraft – the concept was subsequently abandoned.

Figure 23. Yak-7 modification with DM-4S ramjets. Yakovlev

Figure 24. Yak-7 modification with DM-4S ramjets. Yakovlev

Figure 25. Lavochkin La-7 D-10 pulsejet accelerator testbed. Laspace

La-7 D-10 - The unreliability of the LPRE and ramjet accelerator concepts saw focus moved to another type of jet propulsion as an accelerator/booster to increase speed performance of the La-7 piston engine fighter – the pulsating or pulsejet, which had been successfully employed by Germany on the V-1 (Fi-103) flying bomb in the last year of the war. Under chief designer Vladimir N. Chelomey the Soviet Union had commenced work on this type of propulsion at Plant No.51, at the end of the European war in May 1945, data later being gleaned from German pulsejet research. The Soviet research effort resulted in the PuVRD D-10 accelerator, two such units, each rated at 200 kg thrust, to be installed on the La-7, one beneath each wing. Installation work was completed in November 1945 and flight testing commenced in late summer 1946. In regard to the increased maximum speed value, the installation proved successful, the La-7 speed being increased by 119 km/h at an altitude of 3000 m and increased by 193 km/h at an altitude of 800 m in comparison to the unmodified La-7. Development of the concept was continued, but a switch to the La-9 (developed from the La-7, but with a new metal forward section and a laminar flow wing section) was recommended as this design was better suited for development as a potential mixed power fighter.

Despite the switch to the La-9, D-10 pulsejets were installed on three La-7UTI conversion trainer aircraft powered by the AH-82FN. La-7UTI production was undertaken at Plant No.2 (Gorky), 481 units being built between July 1945 (the prototype flew in March 1945) and late 1946. Plant No.163 (Pneza) converted a number of single-seat La-7 fighters to two-seat La-7UTI standard. The installation of the D-10 pulsejet accelerators on the La-7UTI aircraft was not conducted in regard to deep research work, but rather to provide a pool of jet accelerator equipped aircraft to participate in the 1947 Tushino air parade. The La-7UTI D-10 aircraft were put through a basic test phase to clear them for flight operations, but they were dropped from participation in the air parade as the La-9 D-13 combination was ready in time for the parade on 30 August 1947.

Figure 26. La-7UTI without the D-10 installation. Laspace

Further work on jet accelerators for piston engine fighter aircraft led to installation of two PVRD-430 ramjets (designer M.M. Bondaryuk) on the Lavochkin 126 aircraft, which, like the 120, was a prototype development iteration leading to the La-9. The 126 had been flown in December 1945. The major differences from the 120 design was the adoption of the ASh-82FN piston engine, a modified wing with 'plywood sheathing' and a new armament of 4 x NS-23 23 mm cannon. Flight tests of the 126 design continued into April 1946 (pilots A. Davydov, I.E. Fedorov and A Popov) (Laspace).

The 126 was then taken in hand for modification as an experimental mixed power fighter. When the installation of the PVRD-430 ramjets was completed in 1946 the design received the new factory index 164, the aircraft conducting its maiden flight in 164 configuration in June 1946. Factory testing was completed in September that year (the aircraft was test flown by pilots A. Davydov and A. Popov). Speed increase over the unmodified 126 aircraft was 109 km/h '(in relation to the original aircraft without ramjet – 64 km/h)'. The conclusion arrived at from the results of test flights was that the ramjet accelerator was not only more reliable than the LPRE accelerator, but was also less complex in operation.

The success of the tests on the Lavochkin 164 led to the Lavochkin 138, which was a development of the serial La-9 design equipped with PVRD-430 accelerators (the same design as used on the 164) to augment performance. The development road that led to the La-9 saw the design of the Lavochkin 130 that followed the 120 and 126. This design was known within Lavochkin for a time as the 'all-metal' La-7, moving from the partial metal partial wood structure of the 120 and 126 to an all-metal structure. The 130 was to be powered by the ASh-83 piston engine, but, following the decision not to proceed with production of this power plant, it was replaced by the ASh-82FN piston engine. The first Lavochkin 130 was completed in

January 1946 and the factory testing (Lavochkin OKB had returned to Plant No. 301 near Moscow from where it had been evacuated in 1941 due to the German threat to the capital), totaling 30 flights, was completed in May that year. The results had shown the 130 to be superior to not only the La-7, but the Yakovlev Yak-3 and Yak-9U. State tests were successfully completed in October 1946, the design being approved for serial production at Plant No. 21 (Gorky) with the designation of product (type) 48. In service the aircraft was designated La-9. Although almost externally identical, the La-9 could be distinguished from the La-7 by the cut off wing tips of the former, which replaced the rounded wing tips of the latter.

Figure 27. The information that is applied to his photograph is that it is the 138 modification of the Lavochkin 164. Laspace

The 138 development had the specific aim of achieving a speed at least approaching that of the first generation jet fighters, but with the extended range of the piston engine fighter. Two 138 aircraft were flown from February to September 1947. The range, although short of what was desired, was at an acceptable level. However, the speed fell considerably short of what was hoped for. At an altitude of 3000 m the ramjets allowed an increase in speed over the La-9 serial fighter of 45 km/h, which was woefully short of the hoped for speed increase of 70-100 km/h. It was concluded that such performance levels would allow the 138 to be utilised successfully as an interceptor against piston engine bombers of the Boeing B-29/50 class, but, that performance was inadequate to expect more than sporadic success against modern high speed fighter opposition.

Figure 28. La-9 RD-13 pulsejet accelerator test-bed. Laspace

Further work on jet accelerators led to the installation of RD-13 PWUDS accelerators on an La-9 (Lavochkin documentation suggests, but does not confirm, more than one conversion), this design then being designated La-9 RD-13. The aircraft commenced flight testing in this configuration in August 1947 and, on the 3rd of that month, participated in the Tushino air parade. A La-9 RD-13 was transferred from Lavochkin to the Soviet Air Force Research Institute in November 1947 to undertake what were vaguely referred to as 'special flight tests' (Laspace). Overall testing showed the design had a 127 km/h speed increase over the serial La-9 (in relation to the initial aircraft without PWRM – 70 km/h). However, this came at the price of a vast increase in fuel consumption, with its corresponding decrease in range, and unacceptably high 'acoustic loads and vibrations' leading to a recommendation against further work on the concept (Laspace).

The inadequacy and or unreliability of the ramjet, pulsejet and LPRE accelerators for the Soviet mixed fighter concepts led to thoughts on a more practical way to provide piston engine fighter aircraft with additional speed to counter high-speed jet and rocket powered aircraft. This led to the concept referred to as the CJE (Compressor Jet Engine) also referred to as a motor-compressor engine, in which the aircraft tail section would house a jet compressor that would be driven by a shaft rotated by the main piston engine. While Lavochkin had been at the forefront of development of piston engine fighters with rocket or jet accelerators, the bureau was notably absent from work on the CJE concept fighter design or modifications. Other design bureau, Yakovlev, Mikoyan and Sukhoi, were involved in CJE fighter concept studies. The peculiarities of the Sukhoi and Mikoyan designs lay not in any of the design traits themselves, but rather with the fact the designs were formulated from new rather than modifications form any existing design. The ideology behind the rocker/jet accelerator concepts had initially been to decrease the performance disparity between best of the World War II era piston engine fighters and the projected early jet powered fighters – the latter typified by the Me.262. However, by the time development of the Mikoyan and Sukhoi piston/CJE designs were underway the Soviet Union was developing that nations first fighter aircraft powered exclusively by jet propulsion, which would result in the MiG-9 and Yak-15, both of which conducted their respective maiden flights in April 1946. It has to be concluded that development of the Sukhoi and Mikoyan piston/CJE designs was intended as a partial insurance policy against extended delays with the introduction of a first generation jet powered fighter aircraft, Sukhoi developing the Su-9, which borrowed many design traits, including the Jumo-004 engines, from the Me.262, alongside development of the Su-5 piston/CJE and Su-7 piston/rocket accelerator designs. The Yakovlev design continued the wartime trend of adapting existing piston engine fighter designs to be augmented by accelerator power.

Figure 29. Mikoyan I-250 piston/CJE fighter prototype. RAC

The design work on VRDK technology that would lead to the CJE's of 1945 had commenced at TsAGI in 1941 under the supervision of G.N. Abramovich. Failure to bring the various studies to fruition led Abramovich to turn to CIAM, which, under the supervision of K.V. Kholshchevnikov (other prominent figures involved in the development included AA Fadeeva and V.A. Stefanovsky) from mid-1943, took on the task of developing technology to become the motor-compressor in which an axial flow compressor would be combined with a piston engine through a shaft. Design studies led to separate branches – one a petrol (gasoline) fed piston (under K.V. Kholshchevnikov) and the other a diesel fed piston (designer AI Tolstov). The Kholshchevnikov group that went on to design and develop the working motor-compressor (CJE) consisted of V.G. Protserov, N.F. Peshekhonov, N.V. Niktin, G.N. Romanov, P.N. Klimov, V.A. Stefanovsky, A.N. Silin, G.E. Chernenko, V.V. Sokolov, V.A. Vaikov, M.N. Nikolaychik, N.I. Zhidkov and G.G. Myakinkov et al (CIAM).

Mikoyan was authorised to proceed with a piston/CJE (motor-compressor engine) concept through GKO resolution No. 5946 of 22 May 1944 (this called for development by various bureaus of piston-LPRE and piston-jet power plants). The concept would employ the VK-107A (formerly M-107) piston and a VHFK air-jet compressor – CJE developed by K.V. Kholshchevnikov, the design being informally referred to as the Accelerator Kholshchevnikov (CIAM).

Figure 30. The E3020 motor-compressor attached to the VK-107A through a driveshaft. CIAM

The complete power plant consisted of the VK-107A (previously designated M-105) piston engine, which drove the forward airscrew and, by the function of a shaft, via a twin-speed box, drove a single-stage axial compressor jet incorporated in the E3020 jet engine. The rotation speed of the shaft varied according to the altitude, the switch being conducted automatically. The compressor was fed air through an intake located under the hub of the propeller in the aircraft nose. As it left the compressor some of the newly compressed air was directed to the piston-engine blower, which facilitated an increase in altitude by ~1000 m. The major part of the compressed air was directed via a radiator through a combustion chamber (the chamber was cooled through air that flowed through the gap located between the airframe shell and the chamber itself) before exiting through a two-position exhaust nozzle located in the aircraft tail section (CIAM).

The E3020 motor-compressor was not designed to operate throughout the entire flight. It was designed to be employed on take-off to reduce the run and when the aircraft was approaching the maximum speed attainable on the VK-107A piston. The total flight time of use of the CJE would not normally exceed ten minutes. Normal cruise flight and climb to altitude would be accomplished on VK-107A piston only power – 1650 hp. When the CJE was employed this would draw off 280 hp. at the first speed rating and 348 hp. at the second speed rating. The E3020, which had a mass of 140.9 kg, developed a maximum thrust of 344 kgf. Total power output of the piston-CJE equated to 2560 hp. (CIAM). As there is ~3.7 hp. for 1 kgf, this value, it is assumed, takes into account the draw off of 348 hp. from the piston power, although this leaves a slight shortfall value of 15 hp. Fuel consumption was 1235 kg/h (CIAM). This equated to 205.83 kg for the ten minutes of flight specified for the system.

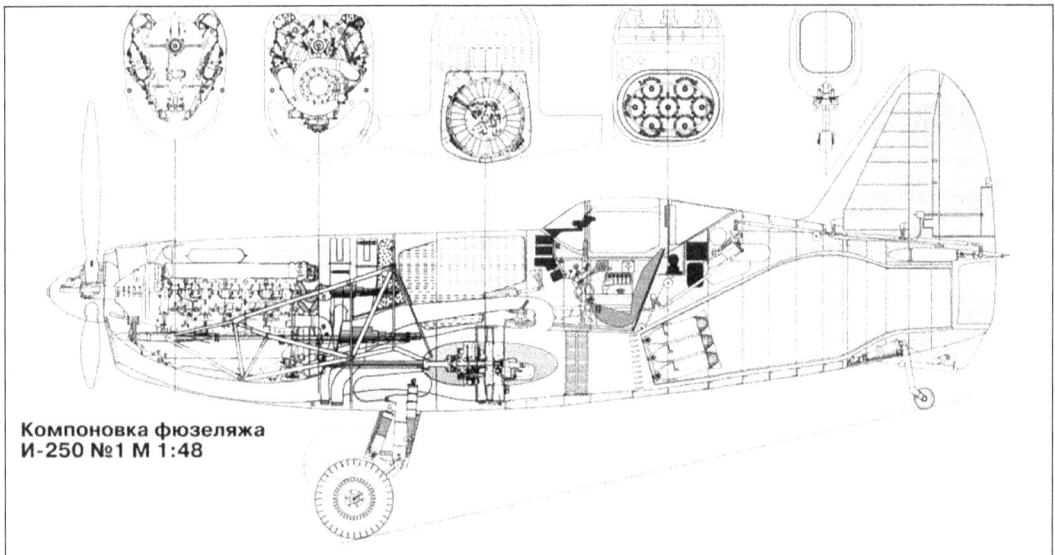

Компоновка фюзеляжа
И-250 №1 M 1:48

Figure 31. Port side on diagram of the Mikoyan I-250 showing the layout of the VK-107A piston/E3020 motor-compressor (CJE). CIAM

The prototype I-250 conducted its maiden flight (pilot A.P. Deyev) on 3 March 1945. The initial flights were conducted under piston power only, but on 8 April 1945 the CJE was activated in-flight. A speed of 809 km/h at an altitude of 6700 m was registered on 13 May 1945 and on 3 July that year this was increased to 820 km/h at the moderately lower altitude of 6600 m. This milestone was significant in not only being the highest speed achieved by the I-250 up to that time, but also the highest speed achieved by any Soviet aircraft up to that time. During this phase of speed increases the prototype was lost when it crashed on 5 July 1945. The I-250 prototype was being tested at maximum speed at low altitude – 650 km/h – when the aircraft crashed as a result of a collapsed stabilizer. The pilot (Deyev) perished after bailing out as the altitude was too low for the parachute to open successfully (CIAM). In the period following the accident the second prototype, which had apparently conducted its maiden flight several weeks prior (FRI pilot A.P. Yakimov), twice had to be landed in an emergency due to in-flight problems. This aircraft incorporated changes to the structure, including an enlarged vertical tail. During these early test flights problems were identified with the supercharger, rectification of which was a collaborative effort involving CIAM. Among the most serious problems encountered was bending of the drive shaft between the piston engine and the compressor, failure of the oil and fuel systems and damage to the compressor blades. The problems were severe enough for K.V. Kholshchevnikov to be summoned to deal with them at the test facility (CIAM).

Figure 32. The I-250 was the first all-metal aircraft designed by Mikoyan. The aircraft was armed with three 20 mm cannon and incorporated armour protection. RAC

Despite the many problems with the aircraft design and the CJE, small-scale pre-series production was authorised to be conducted at Plant No. 381, despite the fact that the aircraft had not yet conducted state tests. To address problems with the life-span of the CJE CIAM endeavoured to bring the period of time that the compressor could be run up to 25 hours initially and then, by April 1946, had increased life-span to 35 hours through modifications to the E3020 axial-compressor and the gearbox. However, problems persisted, including compressor breakdowns, delaying completion of the CJE state tests until May 1947 (CIAM).

The maximum speed attained on factory flight has been unofficially stated as 825 km/h (512.6 mph), this apparently attained at an altitude of 7000 m. However, CIAM records show a maximum achieved speed of 820 km/h achieved at an altitude of 6600 m on 3 July 1945 (CIAM). This later value is the one supported by documentation in the archives.

Figure 33. Mikoyan I.250N three quarters frontal view. CIAM

Figure 34. Mikoyan I.250N three quarters rearward view. CIAM

Although state testing of the I-250 had been delayed into 1947, pre-series production got underway at Plant 381, the Sukhoi Su-5 program suffering due to unavailability of the VK-107A/CJE as a result. The availability of production examples led to the decision not to repair the second prototype after it was damaged during testing. Eight I-250 aircraft had been delivered to the Soviet air force for demonstration by 30 October 1946. Despite the lack of interest by the Soviet air force, these aircraft were acquired for participation in the 1946 Tushino air parade scheduled for 7 November that year – the parade was subsequently cancelled due to adverse weather conditions. The preparations for the parade allowed a short evaluation of the design to be undertaken, the conclusions of which showed the design to be far from satisfactory. On 29 November 1946, N.A. Bulganin, People's Commissar of the aviation industry of the USSR, recommended that of the four experimental designs employing jet propulsion then under review – the MiG-9, Yak-15, La-150 and the I-250 – only the MiG-9, powered by two RD-20 (Soviet copy of the German BMW-109-003A) engines, should be brought into service. The I-250 was discounted as the design was fraught with stability problems, particularly during take-off and. Most importantly, the mixed power concept had lost all relevance now that the first generation jet propulsion fighter aircraft were flying (CIAM). The MiG-9 and, despite Bulganin's recommendation, the Yak-15 progressed to production status as detailed above.

An I-250 aircraft was made available for state testing from 19 September 1947. As there remained no interest from the Soviet air force the program became reliant on Soviet AVMF (naval aviation) for its survival. A series of naval research testing commenced at 'Skulte' airbase, Riga on the Baltic coast of Soviet Latvia on 9 October 1947 and concluded on 3 April 1948 (CIAM). Although the I-250 is often afforded the label MiG-13 during this phase, no record of this designation being applied has been found in the archives (CIAM). There have been a number of claims over the successive seven decades to the effect that the I-250 was introduced to operational service with the AVMF of the Soviet Baltic Fleet, serving for several years. The source of this assertion appears to be the memoirs of A.I. Shahurin who was the Peoples Commissar of the aviation industry of the USSR from 1940-1946. However, despite the high profile nature of the source, this information appears to be erroneous, there being no record in the archives of use of the I-250 beyond the end of the evaluation period in April 1948. In his memoirs, Kholshchevnikov referred to a meeting in which it was stated that an entire Regiment was equipped with the I-250N in the escort fighter/torpedo bomber role. The reaction of the designer of the E3020 was that had no knowledge of the serial production of the compressor – no such serial production, of course, having taken place, the I-250N Regiment in question being fictitious. It can, with confidence, be concluded that the I-250 was operated purely in an evaluation capacity by the AVMF, this ending in April 1948.

There have been several values put forward for the numbers of I-250 aircraft built. Two prototype and eight aircraft as the low and two prototypes and 10+16 aircraft as the high. There are also other values of 12 aircraft thrown into the mix.

Requests for definitive clarification from the modern day parent company of Mikoyan (RAC) has not cleared the haze sufficiently to make a definitive statement on the exact numbers of I-250 aircraft built. However, there is no record in the archives of values beyond two prototypes and a pre-production batch of 12 aircraft.

The retirement from flight can be traced to the last flight of the AVMF evaluation on 3 April 1948. For the purposes of this paper it is accepted that the piston/CJE concept was effectively dead from April 1948. The various design bureau's had halted research and development into such projects as more promising jet propulsion fighter aircraft were already in service and high-performance aircraft such as the Mikoyan MiG-15 and Lavochkin 174 (effectively the prototype of the La-15 interceptor/fighter), with speeds approaching Mach 1, were in flight-test.

Figure 35. Yak-9U serial fighter. MODRF

The Yak-9VRDK conversion can be grouped into the piston/CJE engine mixed power fighter category that included the Sukhoi Su-5 and the Mikoyan I-250. The Yakovlev aircraft, as noted above, differed in that it was a modification of a wartime serial production fighter, whereas Mikoyan and Sukhoi had put forward new design concepts. The Yak-9 had been developed as a single-seat monoplane fighter powered by an M-105 пф piston engine rated at 1,250 hp. The design retained the same wing area as that of the Yak-7, flight weight was 3060 kg and maximum operational speed was 600 km/h, slightly less than that of the Yak-7. The Yak-9 entered serial production in 1942 and from late that year took part in every major operation on the Russian/German fronts. Output was the highest of any Soviet fighter aircraft design during the Great Patriotic War of 1941-1945 – 14,240 units being built.

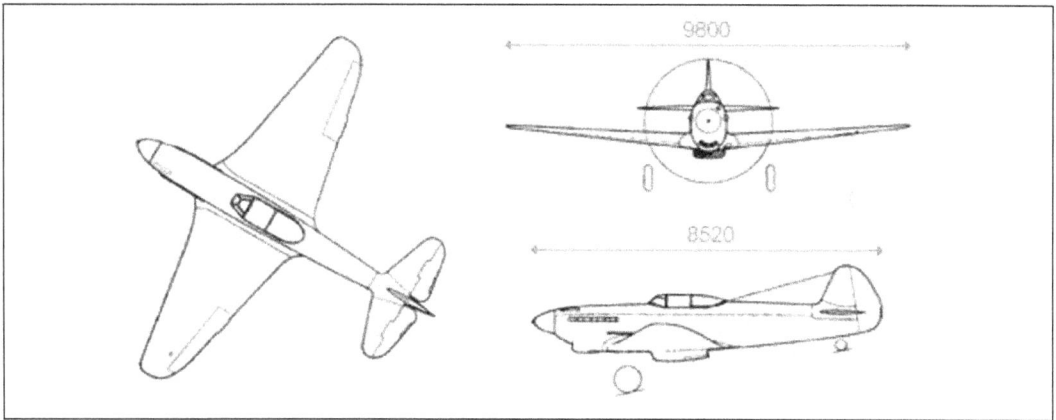

Figure 36. Three view general arrangement of the Yak-9U without the CJE modifications. Yakovlev

Work on the Yak-9 piston/CJE program commenced at TsAGI under the overall direction of the institutes Professor G.M. Abramovich. The piston was a VK-107A (formerly designated M-105) series which was equipped with the necessary modifications to drive the E3020 CJE, which, as was the case with the Mikoyan I-250 and Su-5, reduced the power available to drive the piston. Flight tests of the system showed an increase in attainable speed of around 80 km/h over the unmodified Yak-9, this proving to be inadequate to warrant continuance of work on the program.

Without government backing Sukhoi had commenced preliminary concept studies of an experimental mixed power fighter, the Sukhoi Su-5 (I-107 'D') in January 1944. The design arrived at incorporated a single M-107A (VK-107A) piston engine augmented by a CJE, which in effect held the function of a booster. The engines would be linked through a shaft that would allow the CJE to be rotated by the M-107A piston engine. Sukhoi submitted the design for review by the USSR PCAI leading to it being included in the 1944 draft version of the AFRA (aircraft prototype development plan). The GKS resolution of 22 May 1944 that had authorised the I-250 also called on Sukhoi '… to design and build a single-seat experimental aeroplane with the VK-107A engine, outfitting it with an additional CJE designed and built by TsIAM [CIAM]…' (Sukhoi). This recommendation settled the choice of power plant – the single VK-107A piston engine equipped with a FBDD Fadeev-Kholshchevnikov E3020 CJE (motor-compressor) accelerator that would be employed when additional speed was required in an emergency (Sukhoi).

Detailed design of the experimental fighter commenced at the start of June 1944 under the designation I-107 'D', although this was soon changed to Su-5. The redesign was submitted to be reviewed by the PCAI and the AFRA leading to approval, in autumn of 1944, 'to the opinions delivered on the conceptual design and the report of the mock-up committee' (Sukhoi). The build of the prototype took longer than planned, attributed by Sukhoi to delays in delivery of the power plant.

However, the Su-5 finally got into the air on 6 April 1945 (pilot G.I. Komarov). Factory flight testing continued until the aircraft was involved in an accident, the VK-107A engine breaking up while in flight. There was no replacement engine available. The delay in sourcing a new engine facilitated the time required to provide the Su-5 with a new design of 'laminated-profile wing developed by CAHI [TsAGI]' (Sukhoi). A replacement VK-107A was allocated in early July 1945, and, while it had only a limited service life remaining, it did allow for the recommencement of flight testing from early August that year. However, following a flight on 18 October 1945, it was accepted that the VK-107A service life had been consumed, leading to a halt in flight testing – this effectively bringing about the end of the factory flight test phase. Sukhoi attempts to secure delivery of a further VK-107A engine proved fruitless, there being limited availability of such engines. Unfortunately for Sukhoi, the Mikoyan I-250 CJE fighter program had received priority over Sukhoi for such engines. This design, which Sukhoi acknowledged was 'more advanced' than the Su-5, had, as noted above, entered into small scale pre-series production, making claims on available VK-107A engines (Sukhoi).

Figure 37. Su-5 piston/CJE mixed power fighter. Sukhoi

During the incomplete test program the Su-5 had attained a maximum speed of 793 km/h when the CJE was employed. This, while approaching that of the (still to fly) Yak-15 turbojet powered fighter, which was capable of 805 km/h, was ~30 km/h less than the speed achieved by the rival Mikoyan I-250 and fell well short of the hoped for speeds of other first generation Soviet jet powered aircraft like the MiG-9 (un-flown at that time) which would be credited with a speed of 910 km/h. As further developments in jet powered fighter aircraft were expected to, before long, expand this discrepancy in speed even further, the fact that the Su-5 had not attained the performance requirements set out for the program, combined with the lack of engine availability, led to the November 1946 resolution of the USSR Council of Ministers to end work on the program, the piston engine/CJE concept itself being overtaken by the increasing performance of turbojet powered aircraft.

Sukhoi Su-5 specification

Length: 8510 mm
Wingspan: 10500 mm
Height: 3390 mm
Airborne weight: 3804 kg
Maximum speed at 4350 m altitude with CJE on: 793 km/h
Service ceiling: 11800 m
Time to height altitude, 5000 m: 4.1 minutes
Flight range: 600 km
Take-off run: 345 m
Landing run: 345 m
Engine: 1 x VK-107A piston rated at 1650 hp.
Booster: 1 x TsIAM (CIAM) CJE rated at 300 kg thrust

Table 1. Specification of the Su-5 piston/CJE mixed power fighter concept. Sukhoi

The mixed power fighter concepts had evolved through the war years from their early beginnings as piston engine/LPRE or piston engine/ramjet accelerator designs through to the piston engine/pulsejet and the piston engine/CJE engine designs as the war was ending. The need for faster fighter aircraft, although not removed with the wars end in 1945, was filled by the introduction of the first generation of Soviet jet fighter aircraft, the Yak-15 and the MiG-9. The latter design in particular demonstrated far more promise than even the most advanced mixed power concepts typified by the CJE. Despite the small scale pre-series production of the MiG I-250 for evaluation, the increasing capability of exclusively jet powered fighter aircraft effectively led to the demise of the mixed power concepts of the 1940's – the immediate future of Soviet tactical fighter aviation parity in capability with the western allies of what would become NATO (North Atlantic Treaty Organisation) alliance, which was perceived in the Soviet Union to have replaced Germany as the major threat to the USSR borders, being assured by the introduction of the swept-wing MiG-15 turbojet powered fighter that had flown in prototype form on 30 December 1947. A prototype production plan had been authorised by the Council of Ministers on 11 March 1947 for the Mikoyan design that would utilise the power of the British Rolls Royce Nene engine., this evolving into the MiG-15 design, the performance of which was simply beyond that which could be expected of a mixed power fighter design, such a concept now clearly relegated to obsolescence. The MiG-15 provided the Soviet fighter aircraft forces with a parity with the best available western fighter aircraft that enabled the Communist air forces to not only counter western air power during the Korean war of 1950-1953, but also secured the USSR's western and south western borders against any adventurous moves by the new western alliance building up in western and southern Europe, which was formalised as NATO on 4 April 1949.

APPENDICES

APPENDIX I

Soviet Mixed Power Fighter Designs, 1942-1948 – data from CIAM, Laspace, Sukhoi & Yakovlev			
Type	Piston	Accelerator	Speed
LaGG-3VRD	M-105пф(PF)	2 x VRD-1 ramjet	567-570 km/h[1]
La-7R	AH-82FN	1 x RD-1 LPRE	734 km/h[2]
La 120R	ASh-83	1 x RD-1KhZ LPRE	828 km/h[3]
Su-7	ASh-82FN	1 x RD-1 LPRE	[4]
Yak-3 LPRE	M-105 пф	1 x RD-1KhZ LPRE	782 km/h[5]
Yak-7 Ramjet	M-105 пф	2 x DM-4S Ramjet	665 km/h[6]
Yak-7R	N/A	2 x DM-4S Ramjet/1 x D-1A LPRE	N/A
La-7 D-10	AH-82FN	2 x D-10 Pulsejet	837 km/h[7]
La 164	ASh-82FN	2 x PVRD-430 Ramjet	[8]
La 138	ASh-82FN	2 x PVRD-430 Ramjet	[9]
La-9 D-13	ASh-82FN	2 x D-13 Pulsejet	771 km/h[10]
Yak-9VRDK	VK-105A	FBDD E3020 CJE	680 km/h[11]
MiG I-250	VK-105A	FBDD E3020 CJE	~820 km/h[12]
Su-5	VK-105A	FBDD E3020 CJE	793 km/h

[1] These values assume a speed of 555 km/h for the unmodified serial LaGG-3

[2] This is based on a 90 km/h speed increase over the serial La-7. If the upper value of a 100 km/h increase is used then maximum speed would equate to 744 km/h

[3] This is based on a maximum speed value of 725 km/h for the unmodified Lavochkin 120

[4] No definitive values released

[5] Despite claims that speed was increased by in excess of 182 km/h over the unmodified Yak-3, the speed of 782 km/h would calculate to a true speed increase value of 127 km/h when the maximum speed of 655 km/h of the unmodified aircraft is used as the start point

[6] This value assumes a speed of 613 km/h for the unmodified serial aircraft

[7] This value is calculated on a documented speed increase of 193 km/h at 800 m altitude. At 3000 m altitude the speed increase dropped to 119 km/h

[8] Speed was increased by 109 km/h over the unmodified Lavochkin 126

[9] At 3000 m altitude speed increased by only 45 km/h

[10] Speed increased by 127 km/h over the serial La-9

[11] The value is arrived at with an unmodified start speed of 600 km/h

[12] Based on values released by CIAM

APPENDIX II

German/British/Soviet aircraft powered exclusively by jet or rocket power, 1944-1946 – data furnished through various flight test reports published in Harkins (2013) and the MODRF

Design	Service entry	Maximum speed serial
Me.262	1944	820-850 km/h (~510-528 mph)
Me.163	1944	940 km/h (~584 mph)
Meteor I	1944	660 km/h (410 mph)
Meteor III	1945	837 km/h (520 mph)
Vampire I	1946	869 km/h (540 mph)
Meteor IV	1946	941 km/h (585 mph)
MiG-9	1946	910 km/h (~565 mph)
Yak-23	1946	805 km/h (~500 mph)

APPENDIX III

Soviet Piston/LPRE Mixed Power Fighter Designs – data furnished by Laspace, Sukhoi and Yakovlev

Type	Function	First Flight	Number Built
La-7R	mixed power trials	27 October 1944	2
La 120R	mixed power trials	July 1945	1
Su-7	mixed power trials	11 January 1945	1
Yak-3	mixed power trials	early 1945	1

APPENDIX IV

Soviet Piston/Ramjet Mixed Power Fighter Designs – data furnished by Laspace and Yakovlev

Type	Function	First Flight	Number built
LaGG-3VRD	mixed power trials	5 August 1942	1
Yak-7	mixed power trials	1942	1
Yak-7R	mixed power trials	N/A	N/A
La 164	mixed power trials	June 1946	1
La 138	mixed power trials	February 1947	2

APPENDIX V

Soviet Piston/Pulsejet Mixed Power Fighter Designs – data furnished by Laspace			
Design	Function	First Flight	Number built
La-7 D-10	mixed power trials	late summer 1946	1
La-7UTI	mixed power demo	1947	3
La-9 RD-13	mixed power trials	3 August 1947	1

APPENDIX VI

Soviet Piston/CJE Mixed Power Fighter Designs – data furnished by CIAM, TsAGI and Sukhoi			
Type	Function	First Flight	Number built
Yak-9VRDK[13]	mixed power trials	1945	1
MiG 1.250	mixed power trials/evaluation	3 March 1945	footnote[14]
Su-5	mixed power trials	6 April 1945	1

Note: The Su-7 referred to in this paper is the first sue of this designation by Sukhoi and should not be mistaken for the later Su-7 supersonic ground attack fighter of the 1950's. Sukhoi had no record of the fate of the Su-7 mixed power fighter.

---

[13] This was a loosely applied designation

[14] See main text

GLOSSARY

CAHI	Central Aerodynamic Hydrodynamic Institute
CIAM	Central Institute of Aviation Motors
CJE	Compressor Jet Engine
FRI	Flight Research Institute, Gromov
hp.	Horsepower
kg	Kilogram
Kgf	Kilogram force
Km/h	Kilometre per hour
La	Lavochkin
LaGG	Lavochkin
LPRE	Liquid Propellant Rocket Engine
m	metre
m^2	metres squared
Me	Messerschmitt
MiG	Mikoyan
mm	millimetre
MODRF	Ministry of Defence of the Russian Federation
mph	Miles per hour
NATO	North Atlantic Treaty Organisation
Su	Sukhoi
TsAGI	Central Aerodynamic Institute
TsIAM	Central Institute of Aviation Motors (CIAM)
Yak	Yakovlev
~	Approximately equal to (can also be used to mean asymptotically equal)

ABOUT THE AUTHOR

Hugh Harkins, FRAS is a historian and author with an extensive background in astro/geophysics and studies/research in the wider scientific, aeronautic, astronautic and nautical technical and historical fields. Hugh has published in excess of sixty books; non-fiction and fiction, writing under his given name as well as utilising several pseudonyms. He has also written for several international magazines, whilst his work has been used as reference for many other projects ranging from the aviation industry, international news corporations and film media to encyclopaedias, museum exhibits and the computer gaming industry. Hugh is a member of the Institute of Physics and is an elected Fellow of the Royal Astronomical Society. He currently resides in his native Scotland. Other titles by the author include:

Iskander - Mobile Tactical Aero-Ballistic/Cruise Missile Complex
Orbital/Fractional Orbit Bombardment System - The Soviet Globalnaya Raketa
Counter-Space Defence Co-Orbital Satellite Fighter
Sukhoi T-50/PAK FA - Russia's 5th Generation 'Stealth' Fighter
Sukhoi Su-35S 'Flanker' E - Russia's 4++ Generation Super-Manoeuvrability Fighter
Sukhoi Su-34 'Fullback'
Sukhoi Su-30MKK/MK2/M2 - Russo Kitashiy Striker from Amur
MiG-35/D 'Fulcrum' F – Towards the Fifth Generation
Air War over Syria, Tu-160, Tu-95MS & Tu-22M3 - Cruise Missile and Bombing Strikes on Syria, November 2015-February 2016
Sukhoi Su-27SM(3)/SKM
Russian/Soviet Aircraft Carrier & Carrier Aviation Design & Evolution Volume 1 - Seaplane Carriers, Project 71/72, Graf Zeppelin, Project 1123 ASW Cruiser & Project 1143-1143.4 Heavy Aircraft Carrying Cruiser
Light Battle Cruisers and the Second Battle of Heligoland Bight
British Battlecruisers of World War 1 - Operational Log, July 1914-June 1915
Eurofighter Typhoon - Storm over Europe
Tornado F.2/F.3 Air Defence Variant
Air to Air Missile Directory
North American F-108 Rapier - Mach 3 Interceptor
Convair YB-60 - Fort Worth Overcast
Boeing X-36 Tailless Agility Flight Research Aircraft
X-32 - The Boeing Joint Strike Fighter
X-35 - Progenitor to the F-35 Lightning II
X-45 Uninhabited Combat Air Vehicle
Into The Cauldron - The Lancaster MK.I Daylight Raid on Augsburg
Hurricane IIB Combat Log - 151 Wing RAF, North Russia 1941
RAF Meteor Jet Fighters in World War II, an Operational Log
Typhoon IA/B Combat Log - Operation Jubilee, August 1942
Defiant MK.I Combat Log - Fighter Command, May-September 1940
Blenheim MK.IF Combat Log - Fighter Command Day Fighter Sweeps/Night Interceptions, September 1939 - June 1940
Tomahawk I/II Combat Log - European Theatre, 1941-42
Fortress MK.I Combat Log - Bomber Command High Altitude Bombing Operations, July-September 1941
XF-92 - Convairs Arrow

www.ingramcontent.com/pod-product-compliance
Lightning Source LLC
Chambersburg PA
CBHW061618210326
41520CB00041B/7495